WANTED:
Dr. Cyberg, computer genius.

CRIME:
Treason.

WHEREABOUTS:
Sector 33.

YOUR MISSION:
Find Dr. Cyberg and stop the rebellion on Robot World.

Bantam Books in the
Be An Interplanetary Spy Series

#1 FIND THE KIRILLIAN!
   by Seth McEvoy
   illustrated by Marc Hempel
   and Mark Wheatley

#2 THE GALACTIC PIRATE
   by Seth McEvoy
   illustrated by Marc Hempel
   and Mark Wheatley

#3 ROBOT WORLD
   by Seth McEvoy
   illustrated by Marc Hempel
   and Mark Wheatley

#4 SPACE OLYMPICS
   by Ron Martinez
   illustrated by John Pierard
   and Tom Sutton

BE AN INTERPLANETARY SPY™ ③

ROBOT WORLD

by Seth McEvoy
illustrated by Marc Hempel
and Mark Wheatley

A Byron Preiss Book

BANTAM BOOKS
TORONTO · NEW YORK · LONDON · SYDNEY

To Laure Smith

Seth McEvoy, author, is an active member of the *Science Fiction Writers of America*; a video game designer and programmer; and is currently writing a critical study of the work of Samuel R. Delany.

Marc Hempel and *Mark Wheatley*, illustrators, joined forces in 1980 as Insight Studios to produce comics, illustrations, and graphic design. Marc Hempel has a degree in Painting and Illustration from Northern Illinois University. His work has appeared in *Heavy Metal*, *Epic Illustrated*, *Bop*, *Fantastic Films*, *Video Action*, and *Eclipse*. Mark Wheatley has a degree in Communication Arts and Design from Virginia Commonwealth University. His work has appeared in *Metal*, *Epic Illustrated*, *Zebra Books* and on Avalon Hill Games. Currently he and Marc are collaborating on a graphic story series, *Mars*.

RL *3*, IL age *9* and up

ROBOT WORLD
A Bantam Book/August 1983

*Special thanks to Judy Gitenstein, Laure Smith,
Ron Buehl, Anne Greenberg, Lucy Salvino,
Carol Wheatley, Rick Smith, Ron Bell,
Katheryn Mayer, Claudia Lafufe, Donald Quinn,
John Pierard, Hilary Barta, and Rick Brightfield.*

*Cover art and book design by Marc Hempel
Mechanical production by Insight Studios*

Typesetting by Graphic/Data Services

*"Be An Interplanetary Spy" is a trademark of Byron Preiss
Visual Publications, Inc.*

ISBN 0-553-23700-4

Published simultaneously in the United States and Canada

PRINTED IN THE UNITED STATES OF AMERICA

O   0 9 8 7 6 5 4 3 2 1

Introduction

You are an Interplanetary Spy. You are about to embark on a dangerous mission. On your mission you will face challenges that may result in your death.

You work for the Interplanetary Spy Center, a far-reaching organization devoted to stopping crime and terrorism in the galaxy. While you are on your mission, you will take your orders from the Interplanetary Spy Center. Follow your instructions carefully.

You will be traveling alone on your mission. If you are captured, the Interplanetary Spy Center will not be able to help you. Only your wits and your sharp spy skills will help you reach your goal. Be careful. Keep your eyes open at all times.

If you are ready to meet the challenge of being an Interplanetary Spy, turn to page 1.

TOP SECRET

Enter your
Interplanetary Spy
ISBN number
below.

If you are not sure,
check the back cover of
this book.

Turn to page 2.

You are in a class M starship returning from a mission on the planet Parno when you receive an urgent message from Spy Center.

Red Alert

"This is Agent Tavro. Behind me is a group of colonists from the planet known as Robot World, in Sector 33. They were found shipwrecked in space and seem to have been brainwashed. Using mind probes, we have learned part of their story."

Go on to the next page.

"The colonists' leader was a computer genius named Dr. Cyberg. He and the colonists set up Robot World to see if robots could be used to help settle an uncharted planet.

Dr. Cyberg

"Soon after arriving on Robot World, Dr. Cyberg was injured in an explosion. To save his life, he had to replace one of his eyes and other body parts with machinery. He became part robot himself—a cyborg!"

Turn to page 4.

4 "After he recovered, Dr. Cyberg and the colonists began their experiment. They used robots to make barren lands useful. Cities were built by the robots. Farms were established.

"Then one day a change came over the robots. They stopped following the colonists' orders. When a group of colonists went to the Marox Swamp, where Dr. Cyberg's lab was, the robots would not let them in."

Go on to the next page.

"That's when the robot rebellion began. 'Humans are not perfect,' the robots declared. 'They must be destroyed.' The robots began to attack the colonists.

"Most of the colonists were able to escape in starships, but Dr. Cyberg was nowhere to be found. He is being held prisoner in his own lab.

"Stand by for orders from Spy Center. This is Agent Tavro, beaming out."

Turn to page 6.

6 As Tavro's picture fades, you receive your mission from Spy Center: Travel to Robot World and stop the robot rebellion.

To do this, you must find Dr. Cyberg. He is the key to bringing Robot World back under the colonists' control.

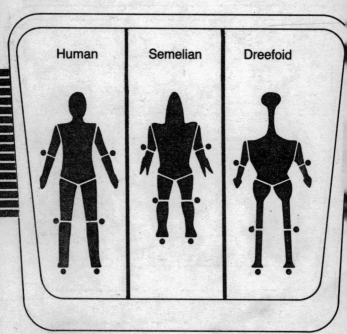

Human   Semelian   Dreefoid

The robots of Robot World will destroy anyone who is not a robot. For your mission, the ship's costuming machine will manufacture a robot disguise.

To begin your mission, find your body pattern among the three shown above.

Go on to the next page.

The outside of your disguise will look like a robot, but the *inside* shape must match the shape of your own body.

The costuming machine can make robot disguises with the following three *inside* shapes:

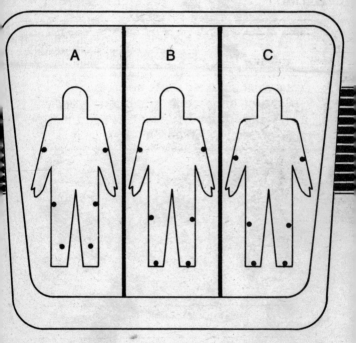

Which one will fit the shape of your body?

Pattern A? Turn to page 10.

Pattern B? Turn to page 12.

Pattern C? Turn to page 16.

Your wrist scanner starts beeping.
The beeps are in code. You listen carefully. It is a message from Spy Center. You must stop your mission to Robot World.

An even greater emergency has come up. You must travel to the other side of the galaxy. The Space Olympics are about to begin. You are needed to stop Gresh, the evil master spy!

Your computer is activated. You must remain in disguise for your entire mission. Your life will be in immediate danger if the robots discover you are human.

To allow you to move around freely, your disguise looks like a general worker robot. Many other robot types were built to handle the different environments of the planet. They were designed for peaceful purposes, but they can be quite dangerous. Below are some of these special robots.

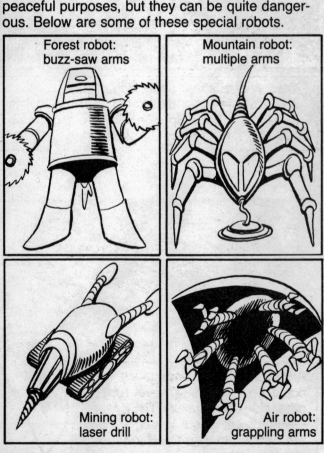

Forest robot:
buzz-saw arms

Mountain robot:
multiple arms

Mining robot:
laser drill

Air robot:
grappling arms

Turn to page 15.

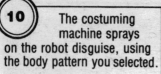

The costuming machine sprays on the robot disguise, using the body pattern you selected.

It doesn't quite fit, but you are in a hurry.

You fly your ship through hyperspace toward Robot World.

Before you can get there, you are attacked by space pirates! You try to fight them, but your disguise is too awkward. You are captured by Mildred and Melissa Khen, wife and daughter of Marko Khen, the Galactic Pirate. They are lowering you into a laser vat. Prepare to become Interplanetary Swiss Cheese!

The End

Everything blows up! You tried your best, but this mission to Robot World was more than you could manage.

The End

Your ship's costuming machine applies the robot disguise to your body.

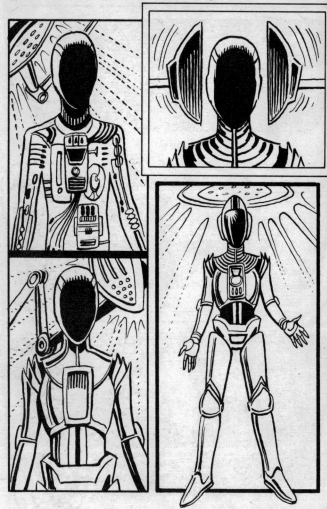

It fits perfectly!

Turn to page 14.

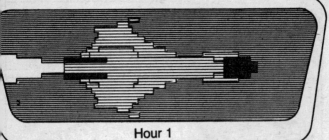

Hour 1

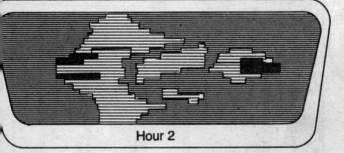

Hour 2

Your ship overheats. You made a fatal error in your calculations.

The ship's costuming machine shows you a scan of your new robot disguise.

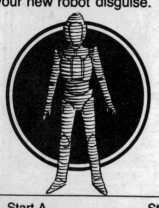

To help you on your mission, your disguise will have a special computer built into it. To activate the computer, you must run your finger along the shortest pathway on its control panel. Which is the shortest pathway?

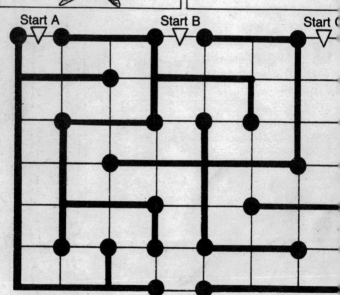

Start A Start B Start C

End Program

Pathway A?
Turn to
page 18.

Pathway B?
Turn to
page 11.

Pathway C?
Turn to
page 9.

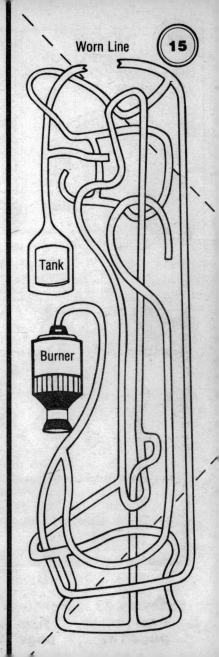

You prepare for travel through hyperspace. But your ship's hyperspace fuel lines are worn out from your trip to Parno. You examine the fuel lines on your computer screen.

You must connect emergency fuel lines so that the fuel can get from the tank to the burner. Fold the top and bottom corners of the page toward you to see your choice of emergency fuel lines available. You must make one unbroken path between the tank and the burner.

Your ship is beginning to heat up from the leaking fuel. If you do not repair the leak in three hours, your ship will explode.

Fold top only? Turn to page 13.

Fold top and bottom? Turn to page 17.

Tank

Burner

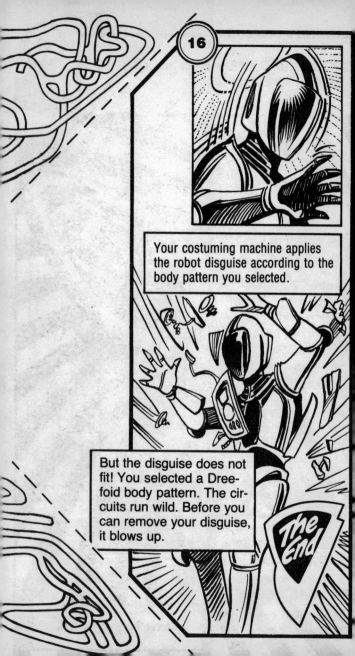

Your costuming machine applies the robot disguise according to the body pattern you selected.

But the disguise does not fit! You selected a Dreefoid body pattern. The circuits run wild. Before you can remove your disguise, it blows up.

The End

You repair your fuel lines and set your course for Sector 33.

Your ship enters hyperspace!

Turn to page 24.

18

Your robot disguise suddenly freezes. You cannot move and you can't get it off. You will be frozen in this position until Spy Center can send another agent.

The End

Excellent! You avoided the pulsar. Your long-range radar detects a small fleet of ships approaching. You get a radio message: "No un-identified spacecraft allowed in Sector 33!"

You've been spotted by the robots—and they have their own rocket ships! You prepare to intercept the ships with your magnetic nets.

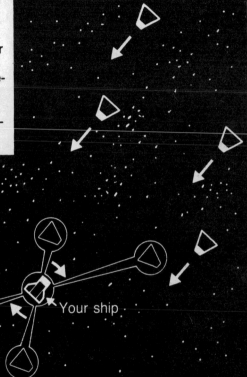

Your ship

You can intercept four ships at one time. You must rotate your ship so that its nets match the shapes of the robot ships. How many quarter turns clockwise will put your nets in the correct position?

 = 1 turn

2 clockwise quarter turns? Turn to page 34.
1 clockwise quarter turn? Turn to page 25.

You track the robot ships into the ion cloud. Your computer warns you that the ion gas is a brainwashing device. Your robot disguise protects you from the effects.

You come out of the ion cloud and see why the colonists were brainwashed!

The robots secretly launched a giant robot ship into space! The last colonists saw it and had to be brainwashed, so they wouldn't be able to tell anyone about it.

It's a space station too! You realize that the robot rebellion is far worse than Tavro reported. Are they planning to take their rebellion to other planets?

Far below the giant robot ship, you see Robot World. You must reach its surface, but how can you get past the giant robot ship? Your magnetic nets are only the size of one of the ship's fingers!

Turn to page 22.

You start to fly past the giant robot when a warning signal comes over your radio. You hear a robot voice say, "Identify yourself."

You contact Spy Center to see if Tavro can get information from the colonists to help you. It will take a few minutes to get an answer, but the giant robot ship won't wait! Its hands start to open. The fingers take off!

Go on to the next page.

You must dodge the flying finger ships before they can blast you!

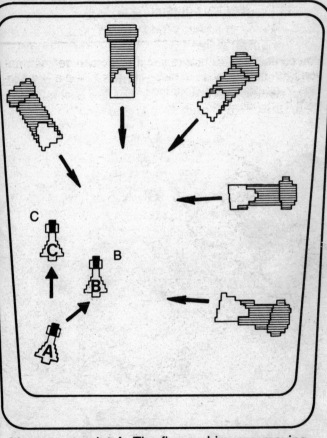

You are at point A. The finger ships are moving toward you. The arrows show the path they will take. Your computer can steer you to point B or point C. To which point should you go?

Point C? Turn to page 33.
Point B? Turn to page 26.

24

Your ship comes out of hyperspace.
You see a pulsar in your path!

The pulsar is a star that expands over a period of
three hours then contracts again. You must get
past it to reach Robot World. Your computer shows
you the pulsar's size during the next three hours.
There are two paths your ship can take to get past
The position of your ship is shown at each hour for
each course. You must keep *outside* the pulsar's
boundary. Which path should you take?

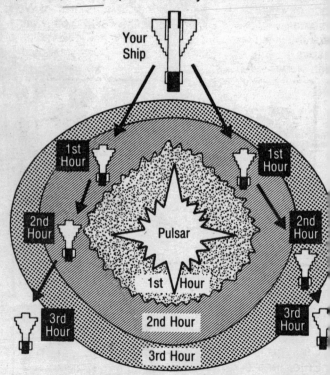

Left path?
Turn to page 13.

Right path?
Turn to page 19.

Good! You rotate your ship correctly and grab four robot rocket ships with your magnetic nets. Then you blast them!

You chase the remaining ships, but they release a strange cloud of ion gas. Your computer indicates that Robot World is somewhere on the other side of that cloud. You must track the ships through the ion gas. You see several shapes on your scanner. They may be space stations, alien worlds, or the robot ships.

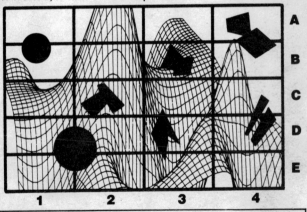

Your computer tracks the robot ships. As a cluster, they have this shape on the screen:

Can you find them in the ion gas? Enter their coordinates:

Are they B-3? Turn to page 20.
Are they C-2? Turn to page 8.

(26)

You move to point B and hope the finger
ships don't hit you.

You blast the finger
ships as they go
by. You get all the
finger ships except
the thumb! You are
blasted!

The End

You send the code word: ◼︎◻︎ ◼︎◻︎ ◼ ◼︎◻︎ ◼︎◻︎ **27**
The missiles return to the giant robot ship.
You are given clearance for landing.

The robots now think you are a robot!
You examine a map of new landing
ports transmitted to your ship by the
robots.

◼︎
| ◼︎◻︎ ◻ ◻︎▬ ◻◻◻

◼︎
◻︎▬ ◼︎◻︎ ◼︎▬ ◼︎◻︎ ◼︎◻︎

◼︎
◻◻◻ ◼︎▬ ◼︎◻︎ | ◻

You pick a port city.

◼︎
| ◼︎◻︎ ◻ ◻︎▬ ◻◻◻

Then you guide
your ship in for a landing.

Turn to page 28.

You descend quickly, but you crash into a waste dump! It looked just like a landing port. Your ship is damaged, but your first concern is to find Dr. Cyberg. You climb out of the hole.

In the distance you see robots building a second giant robot ship!

Go on to the next page.

You must find Dr. Cyberg. You know his laboratory is in the Marox Swamp, but you do not know where that is.

You see many soldier and general worker robots. Luckily for you, your disguise seems to be working. You approach the area where the giant robot is being built. You watch the robots going about their tasks. Some of them are using a special data robot to get information. Perhaps it will work for you.

You saw the robots putting rods across the data robot's top to connect data points. What is the fewest number of rods that will connect all of the data points?

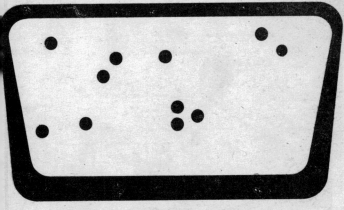

Four rods? Turn to page 41.

Three rods? Turn to page 32.

You try to dodge the missiles. You see that you cannot. They will hit you sooner or later.

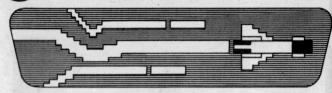

Fortunately your radio message from Tavro arrives! He sends you an alphabet code used by the robots, an identity code, and a code word to stop the missiles. But the ion gas cloud splits the code word! You must shift the bottom parts of the code word to the left to match the top and spell out ROBOT. Quick, before a missile blasts you!

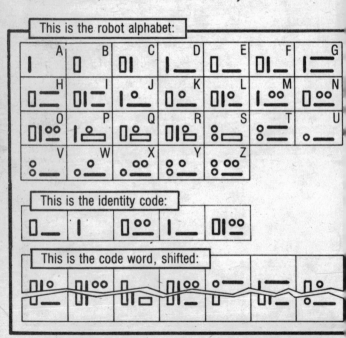

Shift three positions left? Turn to page 11.
Shift two positions left? Turn to page 27.

You examine the detailed map of the Marox Swamp. Swamp rat burrows are marked by ✦

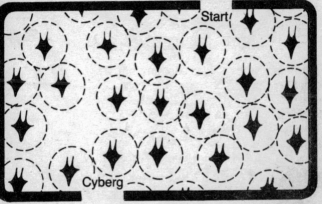

The circles on the map represent the farthest distance a swamp rat will travel from its burrow. You must use your computer to calculate a course through the swamp that will keep you away from swamp rats. Which of the three courses below is safest?

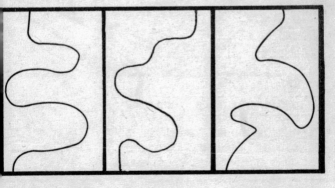

Course A?
Turn to
page 47.

Course B?
Turn to
page 54.

Course C?
Turn to
page 52.

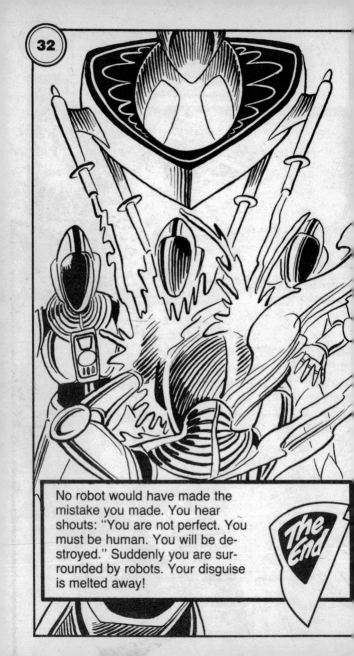

No robot would have made the mistake you made. You hear shouts: "You are not perfect. You must be human. You will be destroyed." Suddenly you are surrounded by robots. Your disguise is melted away!

The End

You manage to dodge all five fingers! As they zoom by, you blast them. Maybe this will be easier than you thought. You fly closer to the giant robot ship.

But as you draw near, you see the shoes of the giant robot opening. Deadly missiles shoot out of the shoes. They are aimed at you.

Turn to page 30.

34

You rotated your ship too far. Your ship's magnetic nets miss the attacking robot ships.

You throw out an emergency force field so they can't blast you, but the Robot World pilots have another trick up their circuits! Since they are not human, they can survive the impact of a crash. They ram their rocket ships through the force field and punch holes in your ship!

The End

You go through the tele-
port station doorway. In-
stantly you are trans-
ported to the Zetan
River.

Before you can go very far, you come to a giant
gate. The gate is made up of . . . robots.

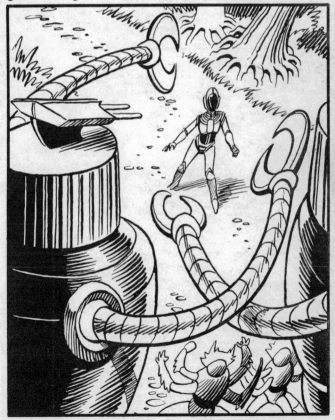

Turn to page 36.

36

The robot gate protects a military base. No robot would be here unless they were part of the army. Before you can escape, you are carried inside the base!

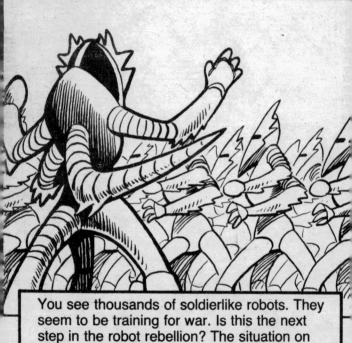

You see thousands of soldierlike robots. They seem to be training for war. Is this the next step in the robot rebellion? The situation on Robot World is worse than you thought.

Go on to the next page.

You are taken to a room in the base for training to be a robot soldier. Your first order is to practice on a video game. If your reflexes aren't as good as a robot's, you will give yourself away. In the game, human and robot shapes come at you. You must shoot the humans but *not* the robots. The robots are training for war . . . against humans!

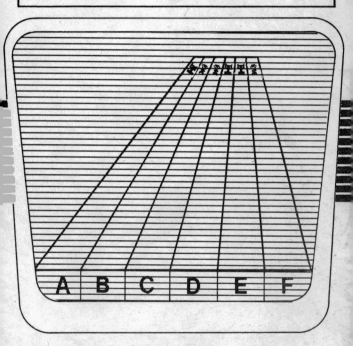

These are robot shapes:

These are human shapes:

Fire at tracks A-D-E? Turn to page 48.
Fire at tracks B-C-F? Turn to page 39.

A bell rings. Your robot reflexes are so quick that you are ordered directly to the base's data center for special training!

Inside, you are left unguarded, and so you try to find out more about Dr. Cyberg. The data center's computer shows you a detailed map of the Marox Swamp. Using your robot disguise's computer, you make a copy of the map.

After dark, you quickly slip outside the data center. You sneak right past the gate robots and head for the Marox Swamp!

Turn to page 40.

40 You arrive at the Marox Swamp. The road ends. You must wade through the swamp.

Luckily your disguise is waterproof. As you descend into the muck, you hear chittering and slurping sounds. Watch your step!

Turn to page 31.

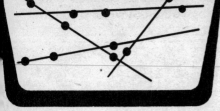

You connect the data points correctly with four rods.

The computer screen lights up. You ask for information about the Marox Swamp.

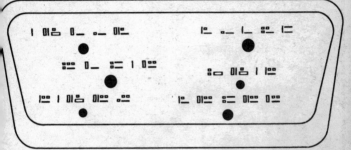

Words appear. They must be names of places. You ask the data robot for clarification.

The data robot tells you to go to the North Teleport Station nearby. The teleport system can take you to the Zetan River Station, near Marox Swamp. The data robot shows you a detailed picture of the Zetan River.

Turn to page 50.

You attach the cooling modules to your suit. You leave the small building.

At first you feel cooler, but then you start to get hot again.

WARNING! HEAT OVERLOAD!

You check the cooling modules, but they have fallen off! You must have taken ones that did not fit properly. You won't have time to go back . . . before you cook!

You arrive at the South Teleport Station.

You must choose the picture that corresponds to the location where you want to go. Do not go to any place more than once.

(Check page 46 if you aren't sure.)

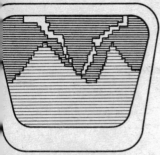

Turn to page 55.

Turn to page 59.

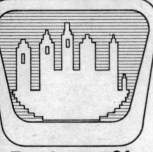

Turn to page 64.

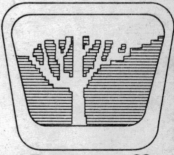

Turn to page 38.

44

You made it past the automatic lasers! You go through a large door and enter what looks like the main control room of Dr. Cyberg's lab. You have not seen any robots in or around the lab, and you suspect that Dr. Cyberg is not here.

You see the control chair that Dr. Cyberg probably used. It is made of floating bubbles. You sit down in the chair. From here you can search the building by remote-control scanner.

Go on to the next page.

The scanner shows you that the entire building is empty. You see a strange button on the control panel. You push it. A probe pricks through your robot disguise to your finger. Then you hear a musical tone and see a video screen coming out of the wall. You see and hear the image of Dr. Cyberg!

"Welcome," he says. "This tape is activated only if a human's blood temperature is detected by my probe button."

So that's why your finger was pricked!
The tape continues: "My house is surrounded by robots. They will capture me at any moment. There is only one robot still loyal to me, ROB-8008. Summon him by pressing the red button." The picture fades. **Turn to page 46.**

46

You press the red button. The face of ROB-8008 appears on the screen. You see his serial number on his chest.

"My master has been kidnapped by robots," says ROB-8008. "I was able to follow him, and I know where he is. But the robots have found my hiding place! They will soon rebuild me into another kind of robot. Then they will probably take me to one of three locations. Use the South Teleport Station to get to them. Hurry!"

You then see pictures of three locations on the screen.
To get to the South Teleport Station, use the special mini-teleport built into the chair. From there, you can travel to any of the three locations to find ROB-8008.

Turn to page 43.

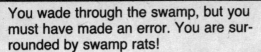

You wade through the swamp, but you must have made an error. You are surrounded by swamp rats!

47

You are outnumbered. Before you are gobbled by the rats, metal hands reach down and lift you up.

You have been rescued by swamp robots. But they have been looking for you! They will take you back to the robot military base. They will make sure you can never escape!

The End

48

You made a mistake! No robot would have done what you did. You are blasted before you can make another move.

The End

You approach the
building and enter. It
is some sort of cool-
ing station, probably
built by Dr. Cyberg.
You search the
building.

Inside you find cooling modules that can be attached to
a robot's metal "skin." The cooling modules will drain off
excess heat. You scan the modules with your computer
and find that only three out of the seven cooling modules will fit your general worker robot disguise.

These are the seven kinds of cooling modules.

A   B   C   D

E   F   G

Which of the modules will fit properly into your disguise's holes?

A-F-G? Turn to page 42.
B-C-E? Turn to page 51.

50

You walk to the North Teleport Station.

From here you can travel instantly to many different locations on Robot World. You must travel to the Zetan River.

(Check page 41 if you don't remember.

Turn to page 35.

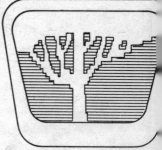

Turn to page 38.

Closed for Repairs

Closed for Repairs

You put on the cooling modules and feel cooler immediately. You continue your journey through the rest of the swamp until you come to a building which is enclosed in a large, slowly rotating bubble. This must be Dr. Cyberg's laboratory!

Turn to page 82.

52

You travel through the swamp, but you wade right into a pack of swamp rats.

Their teeth are sharp enough to bite through your disguise, and they travel faster than you do. Tough luck, Interplanetary Spy!

The End

The laser beams hit your robot suit and bake you like an Interplanetary Pizza!

The End

54 | You make it safely past the swamp rats!

But now you must go through the hottest part of the swamp, an area which is heated by natural radioactivity. You have to get out or you will cook inside your robot disguise.

A computer scan reveals a small building nearby.

Turn to page 49.

You go through the teleporter and arrive at
a mountain. You see lightning striking the
mountaintop every few seconds.

You see several robots taking a gravity tube to the
top of the mountain.

Join them and go to page 56.

You get to the top of the mountain. The thunder is deafening! You follow the robots and see that other robots are strapping them down.

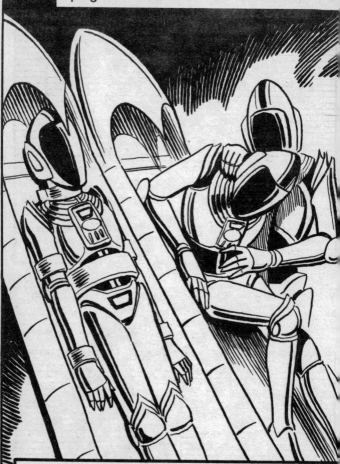

This is the place where old robots go when their electrical systems are worn out. The electrical energy from the lightning will make them young again!

Go on to the next page.

You must pretend to be one of the worn-out robots until you can search for ROB-8008. If he is here, he might be one of the attendants.

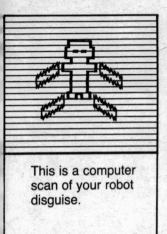

This is a computer scan of your robot disguise.

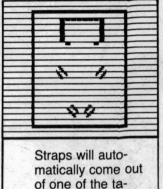

Straps will automatically come out of one of the tables. They will be in this position.

You must choose how to hold your robot body so the straps will *not* touch your body.

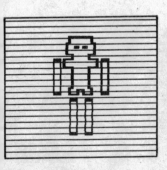

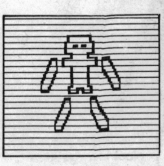

Position A?
Turn to page 11.

Position B?
Turn to page 80.

Good! You get inside the bubble through the moving door. But a scan reveals that there are deadly automatic laser beams guarding the hallways. Your robot disguise can absorb only five of the laser blasts before it breaks down. Which hall should you choose?

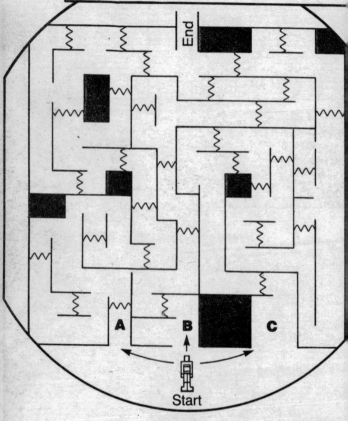

Hall A? Turn to page 53.

Hall B? Turn to page 119

Hall C? Turn to page 44.

You travel through the teleporter and arrive at a volcano! Hot lava flows down one side. You can see a factory built into the other side. It uses volcanic heat to get its energy.

Turn to page 60 and enter the factory.

Inside the factory, fuel cells for the giant orbiting robot are being manufactured. One of the workers may be ROB-8008.

Before anyone gets suspicious, you find a place on the assembly line.

Go on to the next page.

u need a diversion to give you a chance to look ROB-8008. You decide to make a small explo- n in the factory. If you put a fuel cell in the *small* ste dump, it may explode just enough to knock t the robots.

You are working at the top of the factory. If you put a fuel cell into one of the two hoppers, the convey- or belts and other hoppers will carry it to the waste dumps. Careful! If it goes to the big dump, you've had it!

Hopper A? Turn to page 63.
Hopper B? Turn to page 106.

Your computer scans the attacking robot. You discover that it has weak points in its front and back. If you can touch both of its wrists together at the panels on either the front or back, you can stop it cold.

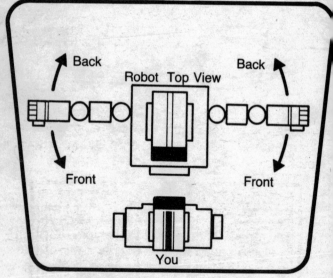

Front? Turn to page 11.
Back? Turn to page 113.

Good work, Spy! The small explosion temporarily knocks out all the robot workers. You examine the serial numbers of all the robots.

None of the robots is ROB-8008. You must try another location by using the South Teleport Station. Search for ROB-8008 at one of these two places, if you have not already been there:

Turn to page 43.

64 You arrive at a floating city. You are on one of the teleportation platforms.

You scan the buildings. They were built for humans. Now they are being used by robots.

You notice one building that seems abandoned. The front door is boarded up. Why wouldn't the robots be using it?

You go around to the back door. An explosion has damaged the barricade. You slip inside.

Turn to page 66.

You get 500,000 points. You have passed the first game with a high score, but it's not high enough to overload ROB-8008's circuits. You proceed to the next level.

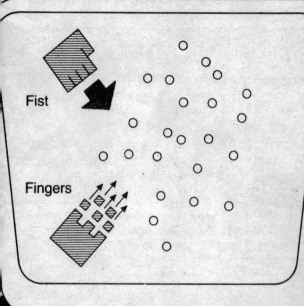

Fist

Fingers

In game two, you must shoot the human ships in the center of the screen. Should you use your entire fist (top), or should you use fingers only (bottom)? This antihuman game is not to your liking!

Fist? Turn to page 11.
Fingers? Turn to page 72.

You enter the abandoned building. There's nothing inside at all. It looks as though the building has not been used since the colonists fled the planet.

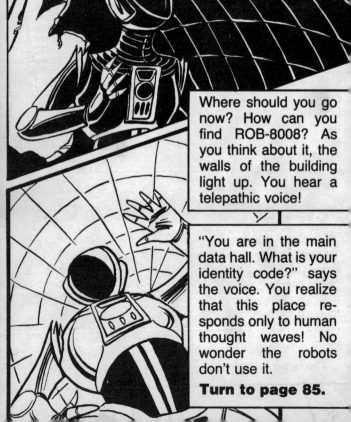

Where should you go now? How can you find ROB-8008? As you think about it, the walls of the building light up. You hear a telepathic voice!

"You are in the main data hall. What is your identity code?" says the voice. You realize that this place responds only to human thought waves! No wonder the robots don't use it.

Turn to page 85.

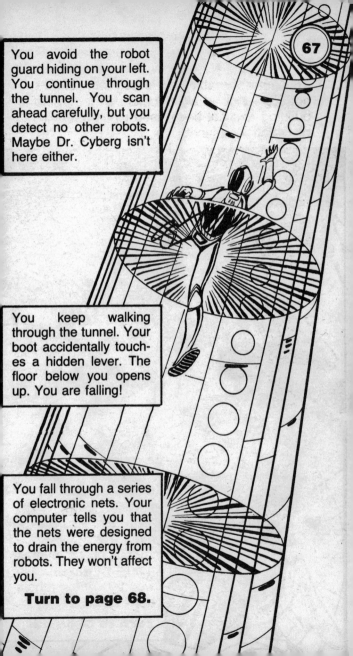

You avoid the robot guard hiding on your left. You continue through the tunnel. You scan ahead carefully, but you detect no other robots. Maybe Dr. Cyberg isn't here either.

You keep walking through the tunnel. Your boot accidentally touches a hidden lever. The floor below you opens up. You are falling!

You fall through a series of electronic nets. Your computer tells you that the nets were designed to drain the energy from robots. They won't affect you.

Turn to page 68.

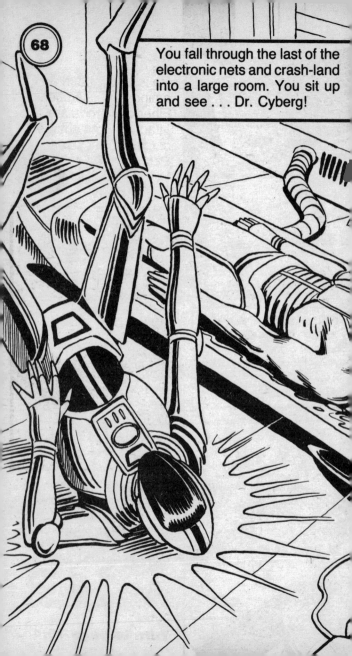

68

You fall through the last of the electronic nets and crash-land into a large room. You sit up and see . . . Dr. Cyberg!

He is enclosed in an amino fluid chamber. Fiber optic rods are attached from a large computer to micro-dishes aimed at his head. He's alive but unconscious.

You go to the control panel of the computer. You must find a way to wake him up. There are no robot guards here. With all humans on the planet gone, they probably didn't figure on anyone trying to rescue Dr. Cyberg.

Turn to page 70.

You examine the computer. Dr. Cyberg's brain waves are being tapped by the microdishes. You must program the computer to use the microdishes to wake up Dr. Cyberg. Here are the computer's circuits for brain wave interference. Connect two matching circuits to cause the microdishes to arouse Dr. Cyberg.

Warning! Incorrect connections may cause the whole computer system to blow up.

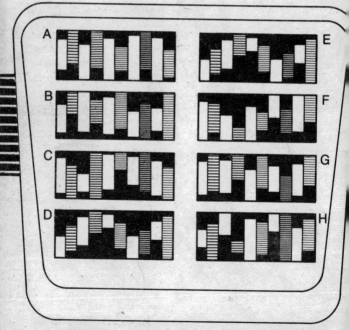

Match A-C? Turn to page 106.
Match D-E? Turn to page 74.

ROB-8008 cannot talk to you in his present form. You must find a way to rewire him so he can tell you where Dr. Cyberg is. Perhaps you can find a way to overload ROB-8008's circuits so that he'll need repairs. Then you can get past his automatic anti-tamper mechanism. You press the start button and see a game based on the giant robot ship.

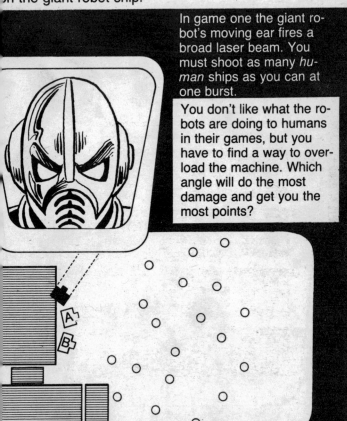

In game one the giant robot's moving ear fires a broad laser beam. You must shoot as many *human* ships as you can at one burst.

You don't like what the robots are doing to humans in their games, but you have to find a way to overload the machine. Which angle will do the most damage and get you the most points?

Ear angle A? Turn to page 110.
Ear angle B? Turn to page 65.

72 You get 1,000,000 points! You are on your way to a top score. There is one more game to play. If you do it well enough, you may be able to overload the machine.

In game three humans have overrun Robot World. Your mission is to protect the last robot family. You can blast the humans one at a time. But after each blast, each remaining human can move. You and the other robots cannot move.

You

Robot Family

Human

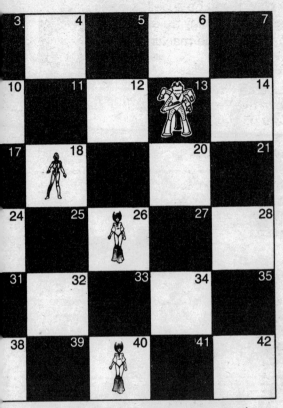

The humans can move one square at a time, diagonally, sideways, up, or down. They will always move toward you or another robot, whichever piece is *closest* to them. If they touch you or another robot, you lose.

You get to shoot first. What is the correct order in which to blast the humans?

Squares 26-8-38? Turn to page 83.
Squares 26-40-1? Turn to page 110.

You did it! The computer uses the microdishes to wake up Dr. Cyberg.

"This robot suit is a disguise," you explain. "I am here from Interplanetary Spy Center to stop the rebellion."

"It's all my fault," answers Dr. Cyberg. "I programmed the robots to make things perfect, but they discovered humans were not perfect. They tried to get rid of all humans to make Robo World perfect."

Go on to the next page.

"They imprisoned me, but most of the other colonists escaped.

"The robots tapped my brain to learn how to run Robot World by themselves. I could not stop them.

"They plan to build a fleet of giant robots to reach other planets. They have already used one to try and stop the last colonists from escaping, but they failed!"

Turn to page 76.

As Dr. Cyberg tries to stand up, you hear a crash! You turn quickly to see robots charging into the room. Dr. Cyberg blasts them with his cybernetic eye. "It sends out a laser to control the robots," he explains, "but it's only strong enough to stun them temporarily.

"We must get to my private scooter dock," says Dr. Cyberg. He activates a secret door. "Follow me," he says, "I know these tunnels better than the robots do!"

You follow Dr. Cyberg through the tunnel.

Go on to the next page.

You go through the tunnels for a short distance. Dr. Cyberg presses a hidden switch. Two power scooters appear. "We must go to my secret laboratory, hidden in the caves. The robots don't know it exists. If we're lucky, they won't be able to follow us."

You get on the power scooters and take off. Each scooter is equipped with its own video screen.

Turn to page 78.

Alarms on the scooter screens go off! "We are coming to an area already protected by robot warning sensors," says Dr. Cyberg. "If we can knock out twenty sensors, their alarm system will be scrambled and we can escape."

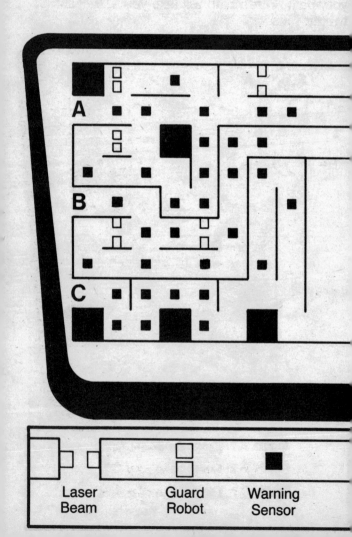

You must use your computer to decide which path will take you to at least twenty warning sensors.

You must be very careful not to cross paths with a robot guard or a laser ray. You cannot retrace your path. Which path will take you to at least twenty sensors?

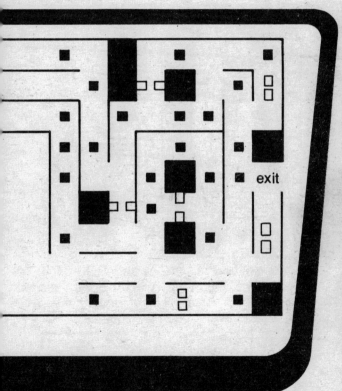

exit

Path A? Turn to page 48.

Path B? Turn to page 86.

Path C? Turn to page 119.

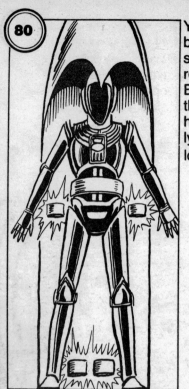

You lie down on the table. The automatic straps cannot quite reach. You're safe! Electricity zaps through the table but does not harm you. While you are lying on the table, you look at all the robots.

None of them has the serial number ROB-8008. You must go back to the South Teleport Station and look for Dr. Cyberg's loyal robot at one of these two places, if you haven't been there before:

Turn to page 43.

⊡_ ⊡▭ ▭_ ⊡_ ⊡▱!

Your password works. The door opens. You go inside and see a long shaft below. It seems to stretch down to the center of the planet!

You wonder why there are no robot guards. There were none at Dr. Cyberg's lab either. Either Dr. Cyberg isn't here, or the robots do not expect any Interplanetary Spies! Since you have no other clues, you begin climbing down the shaft. **Turn to page 96.**

You see no robots guarding the laboratory.
You try to get in, but all the doors are
locked. You scan the bubble with your computer. You get four computer readouts of
the bubble as it rotates.

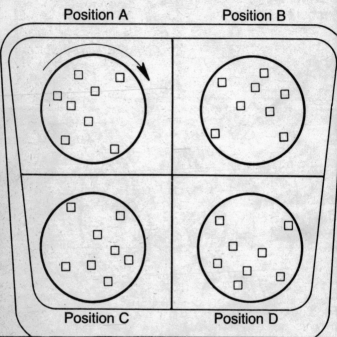

Position A   Position B

Position C   Position D

One of the computer readouts is different. One of
the doors moved! In which position did the door
move? You can go through the moving door to get
inside the bubble. Quick! You can see a laser on
top of the bubble being aimed at you.

Position A? Turn to page 48.

Position B? Turn to page 53.

Position C? Turn to page 58.

Position D? Turn to page 106.

You did it! Your score goes over 10,000,000 points. The machine's circuitry overloads at last! You have done even better than a robot!

Now that the machine is broken, you can repair the circuits and communicate with ROB-8008.

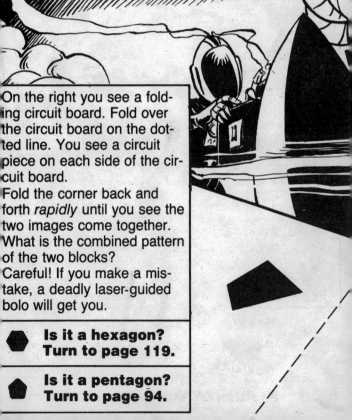

On the right you see a folding circuit board. Fold over the circuit board on the dotted line. You see a circuit piece on each side of the circuit board.

Fold the corner back and forth *rapidly* until you see the two images come together. What is the combined pattern of the two blocks?

Careful! If you make a mistake, a deadly laser-guided bolo will get you.

**Is it a hexagon?
Turn to page 119.**

**Is it a pentagon?
Turn to page 94.**

Before you know what is happening, robot guards surround you! They do not suspect you are a spy. However, since you do not seem to have a job, they will take you to a desert solar energy station.

On the way there, the robot carrier breaks down in the desert. The guards decide they will use *you* as an alternate power source for their carrier. When they plug you into the generator, your disguise will be uncovered. Even if you could escape, it's a long walk home!

The End

ou tell the data hall computer the identity
de that Tavro sent you, which was used
the colonists: ▢_ I ▢▤ I_ ▢▤ .
hen you ask for recent data on ROB-8008.

he data hall responds with the address 107 Tesla
treet. Then it shows you a map of the city. Use the
bot code to read the map. (See page 30.) Which
ay should you go?

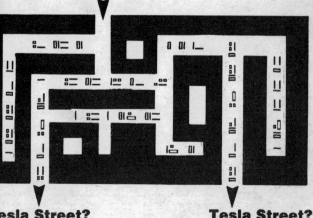

esla Street?
urn to page 108.

**Tesla Street?
Turn to page 32.**

You made it! You managed to knock out at least twenty warning sensors, but a few guard robots have picked up your trail. "My eye is fully recharged now," says Dr. Cyberg. However, instead of blasting the on-coming robots, he directs his cybernetic eye at the cave ceiling.

The guard robots are caught in a cave-in! They will soon dig their way out of it. "The next part of our journey will be very danger-ous," says Dr. Cyberg. "We must go through the Diamond Caverns." **Turn to page 91.**

You arrive at the ice dome. There aren't any guards around, but the entrance is locked.

You notice a keyboard at the door. Enter the password ROB-8008 gave you on page 94.

Is it ☐ ☐ ☐ ☐ ? Turn to page 84.
Is it ☐ ☐ ☐ ☐ ☐ ? Turn to page 81.

"We'll need twenty power crystals to complete our journey," Dr. Cyberg tells you. "No more and no less. The crystals grow in clumps, which cannot be broken apart." You see five caves. Your computer shows you how many crystals are in each cave.

You have time to go to only *three* tunnels, and you can go to a tunnel only *once*. Which *three* tunnels will get you exactly *twenty* power crystals? You must go to the three tunnels you choose to get the crystals. Select your tunnels and proceed!

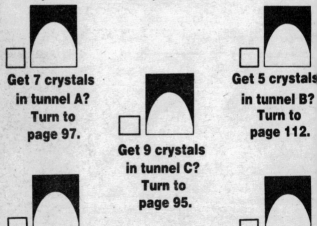

Get 7 crystals in tunnel A?
Turn to page 97.

Get 5 crystals in tunnel B?
Turn to page 112.

Get 9 crystals in tunnel C?
Turn to page 95.

Get 4 crystals in tunnel D?
Turn to page 116.

Get 3 crystals in tunnel E?
Turn to page 93.

Good! You use four crates to trap the robots. Even their laser drills cannot break through the supersteel crate walls. Dr. Cyberg rushes over to the elevator and inserts the four remaining power crystals into a small hopper.

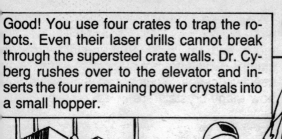

The elevator door opens. But when you look inside, the elevator cable is broken! "It is the only way to my secret laboratory," says Dr. Cyberg. He looks up and gasps.

Turn to page 90.

A flying robot is coming right at you! How can you escape?

Dr. Cyberg uses his cybernetic eye to bring the flying robot under control.

You and Dr. Cyberg grab on to the flying robot and ride it down the elevator shaft.

Turn to page 9

The Diamond Caverns' walls are very sharp.

You must prepare for your trip through the cavern. Trace the path you will be taking on your scooter's video screen without touching a top or bottom wall.

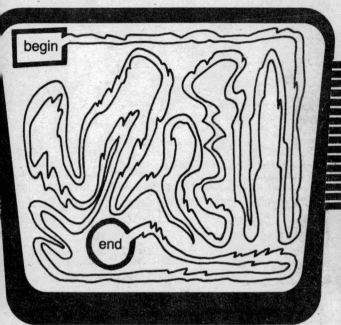

Once you have traced your path, you must set your scooter's autodrive pattern. Which of the three patterns below will match your path through the tunnels?

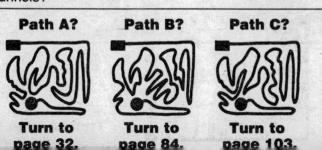

Path A?	**Path B?**	**Path C?**
Turn to page 32.	Turn to page 84.	Turn to page 103.

92 The giant robot's head will separate from the body at any moment. You seal off the airtight doors in the giant robot's neck. You make sure that Dr. Cyberg is all right.

"Even though I know no one will blame me, I feel responsible for what happened on Robot World," says Dr. Cyberg. "I told my robots to be perfect, and they obeyed my orders, perfectly!"

Turn to page 121.

You are attacked by soldier robots. You don't have to look for the power crystals. The soldier robots have them. They give them to you. . . right between the eyes!

94

Excellent. You see a pentagon! You are able to repair ROB-8008's circuits. He cannot speak, but at least he can show you pictures.

A picture of an ice dome appears on ROB-8008's video screen. Then a password in robot code appears on the screen. This is the password:

0_ 0°° :_ 0_ 0│≐.

You leave ROB-8008 and go to a nearby information terminal. You consult a map and travel to the ice dome. It is not far away.

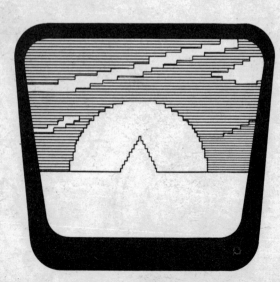

Travel to page 87.

95

You enter the tunnel. A giant cave lizard attacks you! Dr. Cyberg blasts the lizard with his cybernetic eye.

You get 9 power crystals. If you have 20, turn to page 104. If not, return to page 88.

Your descent takes over an hour. You come to another tunnel at the bottom. You walk along it.

Your computer warns you that there may be a hidden robot nearby. It could be to your right or left. Go the opposite way from the robot!

Go left? Turn to page 11.
Go right? Turn to page 67.

You enter the tunnel. Soon you come to an underground river. You almost fall in, but Dr. Cyberg saves you!

You get 7 power crystals. If you have 20, turn to page 104. If not, return to page 88.

The flying robot drops you off at the bottom of the shaft.

"This way!" says Dr. Cyberg. "We're almost there!" You hear a grinding noise, but you keep running.

Suddenly robots burst through the tunnel walls!

You run for it!

Go on to the next page.

You are surrounded by robots, but Dr. Cyberg seems unconcerned. He is examining part of the tunnel wall.

"We've made it!" says Dr. Cyberg. "Here is my secret laboratory, just as I left it!"

You turn and see nothing but a small steel plate in the rock wall. There is no door. Has Dr. Cyberg's mind snapped under the strain?

He blasts the steel plate with his cybernetic eye.

"No robot could ever get in here," he says.

The steel plate moves aside, revealing a tiny room, no bigger than your head!

Turn to page 100.

Dr. Cyberg smiles and turns his cybernetic eye on *you!* A beam of ruby light strikes you, but it does not hurt. The light bounces back and hits Dr. Cyberg.

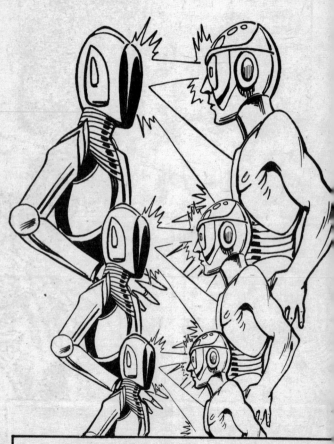

You and Dr. Cyberg start to shrink! He jumps into the tiny room behind the steel plate and says, "Follow me!"

Go on to the next page.

"We'll be safe in here," says Dr. Cyberg. "Not even my robots can drill through super-steel."

Just before the door shuts, one of the robots manages to hit Dr. Cyberg with a stun ray. "You must stop all the robots!" says Dr. Cyberg. "Listen closely. I was designing an emergency robot freeze ray just before the rebellion started. The ray will shut down all robots at once." Dr. Cyberg slumps to the floor. **Turn to page 102.**

It's up to you! You study Dr. Cyberg's equipment. By correctly placing circuit tiles on the control board, you can activate a chain reaction that will set off the robot freeze ray. This is what the circuit shape should be:

Here are the different kinds of tiles which can guide the chain reaction. You can turn these tiles in any direction.

A   B   C   D

Put the chain reaction tiles here:

1	2	3	4	5
6	7	8	9	10

Enter the letters of the patterns here, in the order (1-10) that you put them above:

Is it CDCCCBDBDB? Turn to page 111

Is it ACDCACDADC? Turn to page 114

You make it through the Diamond Caverns with inches to spare. You come out and see a bottomless pit ahead!

Your scooters are able to jump over the pit, but all of your remaining power is used up. You can go no farther. Dr. Cyberg scans the area with his cybernetic eye.

Turn to page 88.

You get back to the scooters with the twent power crystals. Dr. Cyberg puts nine in on
and you put seven in the other. "We will nee
the other four to get into my secret lab," h
says.

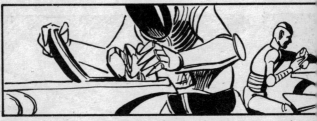

You jump on your fully charged scooters and rac
through the tunnels again.

Soon you arrive in a cavern which has an elevato
that can take you down to Dr. Cyberg's secret lab
But mining robots are guarding the elevator! Ther
are too many for Dr. Cyberg's eye to control.

Turn to page 107.

You arrive inside the giant robot ship safely, but there's a robot waiting to greet you! "Be careful," says Dr. Cyberg. "It's a suicide robot. If we don't stop it in fifteen seconds, it will blow up!"

Before he can use his cybernetic eye, the robot knocks Dr. Cyberg to the floor. You must deactivate the robot by using your hands!

Turn to page 62.

You must stop the mining robots so you can get to the elevator. You notice that there are large storage crates made of super-steel hanging above your heads. You can drop the crates on the mining robots and trap them.

What is the fewest number of crates that will contain the mining robots? You may tilt the crates to cover them.

This is the size of one box:

Mining robot: ●

4 crates?
Turn to page 89.

3 crates?
Turn to page 53.

You arrive at 107 Tesla Street. You go inside and discover an arcade! The robots here are using games to sharpen their skills, just like the robots at the military base.

You look at the machines. The biggest of them resembles ROB-8008! His robot body has been converted into a video game.

Turn to page 7

You ask Dr. Cyberg about the giant robot ship orbiting around the planet. Did the robot freeze ray reach it? Dr. Cyberg uses the deep-space scanner.

"No," says Dr. Cyberg. "The giant robot is too far out in space to be affected by the ray. We've got to stop it before it can escape to another sector!"

Turn to page 118.

The robots of Robot World play video games for keeps! Video games are used to weed out the unfit robots. "Imperfect!" says the machine. It's ROB-8008's voice!

The video game zaps you with enough voltage to power a small starship!

The End

You put the tiles down on the control board. The chain reaction begins.

But the pattern is not quite balanced. The chain reaction back-fires. Instead of freezing the robots, it freezes you. You can't move or talk or even think. You will thaw out in 99 years!

The End

You enter the tunnel. Suddenly you are attacked by mining robots. There is no escape. You don't get the five power crystals. However, the mining robots get you. You find their laser drills very *boring!*

The End

You put the robot's wrists together behind its back. The robot slumps to the floor.

Dr. Cyberg is beginning to come to. You go to his aid. "I'm all right," he says. "Now we must get to the main control room. It's in the head section."

You help Dr. Cyberg through the corridors of the giant robot's body. Alarms are ringing. More robots will be coming. **Turn to page 120.**

You did it! The chain reaction is successful. The robot freeze ray is activated. You use Dr. Cyberg's laboratory scanner to see what is happening on Robot World's surface. All the robots of Robot World have been frozen!

All robot action on the surface of the planet has stopped! The robot rebellion has been ended.

Dr. Cyberg stirs. You help him to his feet and show him what has happened. "If only I could have had a chance to use the freeze ray before the rebellion started!" he says. "The robots will remain frozen. When we return, the robots will be harmless."

Turn to page 109.

You enter the tunnel. You come to an underground river. While you search for crystals, an octorobot attacks Dr. Cyberg. Dr. Cyberg's cybernetic eye is charged up again. He keeps the octo-robot away by using his eye.

You get 4 power crystals. If you have 20, turn to page 104. If not, return to page 88.

You did it! You blocked the correct fiber optic cables. The parts of the giant robot ship cannot communicate with the head. The giant robot ship starts to break up. Only the head will survive.

Turn to page 92.

"We must use my private teleport machine to reach the giant robot ship," says Dr. Cyberg. "It will be 100,000 kad-miles out in space in another hour!"

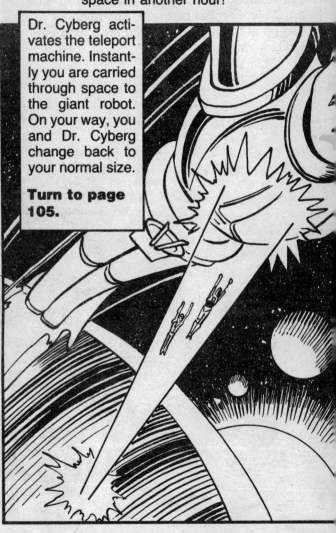

Dr. Cyberg activates the teleport machine. Instantly you are carried through space to the giant robot. On your way, you and Dr. Cyberg change back to your normal size.

Turn to page 105.

You did not choose correctly! The laser beams are guides for bolos. Before you can escape, the bolo's steel wires wrap around your robot disguise. You're all tied up at the moment ... and for a long time to come.

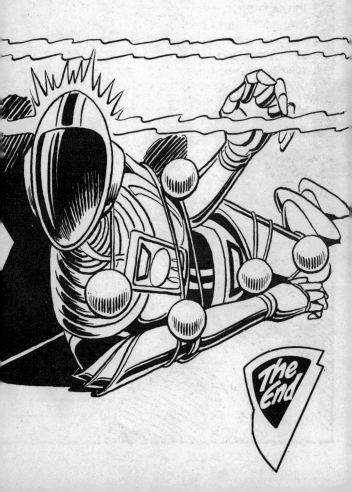

The End

You arrive at the control room. Dr. Cyberg seals it off and activates the main readout screen. It shows all the parts of the giant robot. "They tapped my brain to design this ship," says Dr Cyberg. "I know all its weaknesses."

Directions are sent by fiber optic cables from the head of the giant robot to its other parts. You must block off the cables so that no information can pass from the head to the hands and feet. Without directions, the other parts of the ship will malfunction and break up. You have ten blockers to put at the numbered positions. Where should you put them?

Hurry! The giant robot ship's defense lasers will activate in fifteen seconds. You only have ten blockers to use.

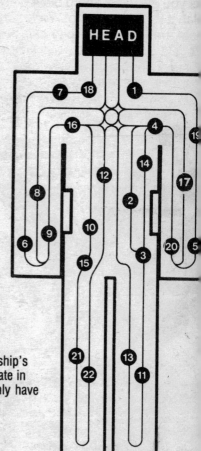

These: 18, 8, 9, 13, 14, 1, 22, 17, 15, 20? Turn to page 53.

These: 1, 17, 20, 3, 13, 12, 6, 8, 16, 10? Turn to page 117.

You radio Spy Center and tell them that you and Dr. Cyberg have stopped the robot rebellion!

As you and Dr. Cyberg blast off, you look back at the planet known as Robot World. Dr. Cyberg says, "Robots are useful, but they can never replace people. People are imperfect, but they are better than machines. People think and feel. They are alive!"

You determine the coordinates of Spy Center and accelerate into hyperspace. Dr. Cyberg will be reunited with the other colonists. They will be able to return safely to Robot World. Your mission is a success, Spy! You have done well.

The End

of this mission.

If you enjoyed this book, you can look forward to these other **B**
An Interplanetary Spy books:

#1 FIND THE KIRILLIAN! by McEvoy, Hempel an
Wheatley

The ruthless interplanetary criminal Phatax has kidnapped Princ
Quizon of Alvare, guardian of the Royal Jewels. You must journe
to the planet Threefax, find the Prince and capture Phatax!

#2 THE GALACTIC PIRATE by McEvoy, Hempel an
Wheatley

Marko Khen, the Galactic Pirate, has been using his band of crin
nals to kidnap rare animals from the Interplanetary Zoo. You mu
find Marko Khen and prevent him from changing the animals in
monsters.

#3 ROBOT WORLD by McEvoy, Hempel and
Wheatley

Dr. Cyberg, the computer genius, has designed a planet of robo
to help humanity. But the robots rebel and Dr. Cyberg disappear
You must track down Dr. Cyberg and face one of the most incred
ble starships in the galaxy!

#4 SPACE OLYMPICS by Martinez, Pierard and Sutt

The insidious Gresh, master spy, has threatened to sabotage th
galaxy's most famous sports event. You must protect the star
the planet Nez, the superathlete Andromeda, as she makes h
way through the games of the Space Olympics!